Abdelhafid Mimouni

Cobalamin Synthesis Protocol

AF294829

Abdelhafid Mimouni

Cobalamin Synthesis Protocol

ScienciaScripts

Imprint

Any brand names and product names mentioned in this book are subject to trademark, brand or patent protection and are trademarks or registered trademarks of their respective holders. The use of brand names, product names, common names, trade names, product descriptions etc. even without a particular marking in this work is in no way to be construed to mean that such names may be regarded as unrestricted in respect of trademark and brand protection legislation and could thus be used by anyone.

Cover image: www.ingimage.com

This book is a translation from the original published under ISBN 978-3-330-87739-9.

Publisher:
Sciencia Scripts
is a trademark of
Dodo Books Indian Ocean Ltd. and OmniScriptum S.R.L publishing group

120 High Road, East Finchley, London, N2 9ED, United Kingdom
Str. Armeneasca 28/1, office 1, Chisinau MD-2012, Republic of Moldova, Europe
Printed at: see last page
ISBN: 978-620-3-22153-4

Copyright © Abdelhafid Mimouni
Copyright © 2024 Dodo Books Indian Ocean Ltd. and OmniScriptum S.R.L publishing group

Cobalamin Synthesis Protocol

Author : Dr. Abdelhafid Mimouni

An independent researcher in bioinorganic chemistry, Dr Mimouni is an expert in macromolecular synthesis and characterisation. He obtained his PhD in Chemistry from the University of Paris XII in 1997, after a Diplôme des Études Approfondies in Bioinorganic Systems from the University of Paris XI in 1993, where he also obtained his Licence and Maîtrise in Chemistry.

Summary: This book examines the cobalamins, in particular hydroxocobalamin, ethylcobalamin and glutathionecobalamin, highlighting their structure, synthesis and applications. Cobalamins are essential to health, playing a key role in DNA synthesis and the formation of red blood cells. Detailed synthesis protocols guarantee efficient methods for their production. In medicine, they are used to treat deficiencies and as antidotes, while in research they serve as valuable tools for studying various biological processes. Analysis techniques such as microanalysis and UV-visible spectroscopy ensure product quality. Future prospects in this field promise to enrich our understanding of diseases and improve treatments.

Book outline: Cobalamin synthesis protocol

Introduction

Introduction to vitamin B12 and its forms

Vitamin B12, also known as cobalamin, is an essential water-soluble vitamin that plays a fundamental role in a number of biological functions. This vitamin is unique, not only because of its complex chemical structure, but also because of its various forms, notably hydroxocobalamin, methylcobalamin and ethylcobalamin. Each of these forms has distinct characteristics and specific applications in the human metabolism. For example, hydroxocobalamin is often used to treat vitamin B12 deficiency, thanks to its ability to transform into other active forms in the body. Methylcobalamin, for its part, is directly involved in the synthesis of neurotransmitters and the protection of nerve cells, while ethylcobalamin is being studied for its potential effects on cellular health and its role in the management of neurodegenerative diseases.

History of cobalamines

The history of vitamin B12 dates back to the beginning of the 20th century, a period marked by significant advances in the field of medicine. It was at this time that scientists began to explore the causes of pernicious anaemia, a serious disease characterised by a lack of red blood cells. In 1926, George Whipple and his colleagues demonstrated that a dietary factor, contained in the liver, could correct this anaemia in animals. This pioneering discovery paved the way for the identification of vitamin B12 as the main factor responsible for the regeneration of red blood cells. In the 1940s, several researchers, including the famous American biochemist, succeeded in isolating and crystallising vitamin B12, thus revealing its unique properties. In 1956, Dorothy Crowfoot Hodgkin succeeded in determining the crystalline structure of vitamin B12 using X-ray diffraction, a breakthrough that not only deepened our understanding of the vitamin's chemical composition, but also laid the foundations for further research into its various forms and their functions.

Importance of cobalamins in biology

Cobalamins are crucial to the proper functioning of cell metabolism. In particular, they are involved in DNA synthesis, a vital process for cell division and tissue regeneration. Their role in the formation of red blood cells is also decisive, as they help prevent conditions such as megaloblastic anaemia, a condition characterised by the production of abnormal red blood cells. In addition, cobalamins are involved in the metabolism of fatty acids and in energy production within cells, thus contributing to the body's overall health. In short, cobalamins are essential cofactors that support many essential physiological functions. Understanding and synthesising them is a subject of great importance in biomedical research, especially as vitamin B12 deficiency can have serious consequences for health, including

neurological disorders and cognitive problems.

Chapter 1: Theoretical basis

1.1 Structure and function of cobalamines

Cobalamins, or vitamins B12, are organometallic complexes containing a cobalt ion at the centre of a tetrahedral ring. The basic structure of cobalamins is composed of a corrin core, which is similar to the structure of haem, but with key variations. This core is made up of four linked pyrrolic rings.

The cobalt in cobalamins can adopt different oxidation states, and is able to bind to various functional groups, including the methyl group, the adenosyl group and the hydroxyl group, giving cobalamins functional diversity.

In terms of function, cobalamins play an essential role in a number of biological processes, including :

- **DNA synthesis**: Cobalamins are involved in the synthesis of nucleic acids, notably through the metabolism of homocysteine, a reaction essential for the production of methionine and, by extension, for protein synthesis.
- **Red blood cell formation**: Crucial for the maturation of erythrocytes, preventing anaemia.
- **Neurological function**: Cobalamins contribute to the myelination of neurons, playing a key role in maintaining a healthy nervous system.

1.2 Chemical reactions of cobalamines

Cobalamins are reactive molecules which take part in various chemical reactions in the body s. The main reactions include :

• **Methylation reactions**: Cobalamins act as cofactors in methylation reactions, which are essential for the transfer of methyl groups, thus influencing the regulation of gene expression and amino acid metabolism.

• **Decarboxylation reactions**: Certain forms of cobalamins, such as adenosylcobalamin, are involved in the decarboxylation of fatty acids and amino acids, helping to produce energy.

• **Redox reactions**: Cobalamins can also act as reducing agents in various redox reactions, helping to maintain cellular redox balance.

These reactions illustrate the importance of cobalamins as enzymatic cofactors, playing a crucial role in cell metabolism and the synthesis of essential biological molecules.

1.3 Importance of the inert atmosphere in organic chemistry

In organic chemistry, the handling and synthesis of compounds sensitive to oxidation or humidity often require inert atmosphere conditions. The use of an inert atmosphere, generally consisting of argon or nitrogen, is essential for several reasons:

- **Preventing oxidation**: Many reagents and intermediates in organic chemistry are sensitive to oxygen, which can lead to undesirable reactions or degradation of the compounds. Working in an inert atmosphere minimises these risks.

- **Reagent stability**: Some compounds can be hygroscopic or react with water. By avoiding humidity, the stability of the reagents is preserved, guaranteeing reliable experimental results.

- **Improved yields**: By eliminating side reactions caused by oxygen or moisture, the yields of organic syntheses can be significantly improved, which is crucial in large-scale production processes.

- **Facilitation of sensitive reactions**: Some reactions, such as chemical reductions or couplings, require strict conditions to be effective. The use of an inert atmosphere creates an optimal environment for these reactions.

In conclusion, understanding the structure and functions of cobalamins, their chemical reactions, and the importance of the inert atmosphere is essential to appreciate their role in biology and their application in organic chemistry. These theoretical foundations provide the necessary framework to explore the syntheses and applications of cobalamines in depth.

Chapter 2: Synthesis of ethylcobalamin

2.1 Introduction to ethylcobalamin

Ethylcobalamin is a valuable derivative of vitamin B12, characterised by the addition of an ethyl group (C_2H_5) to the cobalamin ring. This structural modification gives it unique properties, making it a subject of growing interest in the fields of biology and pharmacology.

The metabolic properties of ethylcobalamin are promising, particularly for its potential in the treatment of certain vitamin B12 deficiencies. Indeed, studies suggest that ethylcobalamin may enhance the absorption of vitamin B12 by cells, thereby increasing its therapeutic efficacy. Research into ethylcobalamin is also extending into the development of new therapeutic agents, particularly in the context of neurodegenerative diseases, where vitamin B12 derivatives could play a protective role against neuronal degradation.

2.2 Materials and reagents required

The following materials are needed to synthesise ethylcobalamin:

- **Vitamin B12**: 250 mg (or 0.0005 moles), serving as the main substrate for synthesis.

- **NaBH$_4$ (sodium borohydride)**: 18.9 mg (or 0.0005 moles), used to reduce vitamin B12.

- **Ethyl iodide (C_2H_5I)**: 150 µL (approximately 0.0015 moles), key reagent for cobalamin ethylation.

- **Distilled water**: for dissolving reagents.

- **Ethanol or methanol**: as a solvent for ethylation, promoting solubility of the reagents.

- **Dilute acid (e.g. 0.1 M HCl)**: for neutralising sodium borohydride after reduction.

- **Laboratory equipment**: beakers, test tubes, pipettes, fume hood and other equipment needed to ensure a safe and controlled working environment.

2.3. Protocol for the reduction of B12 with $NaBH_4$

1. **Dissolution of Vitamin B12 :**

o   Dissolve 250 mg of vitamin B12 in 10 mL of distilled water to obtain a solution with a concentration of 25 mg/mL. This dissolution ensures even distribution of the reagents.

2. **Calculation of $NaBH_4$:**

o   For each mole of vitamin B12, approximately one mole of $NaBH_4$ is required.

o   **Calculation:** 0.0005 moles of B12 $\times$ 37.83 g/mol $\approx$ 18.9 mg of $NaBH_4$.

3. **Reduction reaction :**

o   **Preparation**: Place the vitamin B12 solution in a suitable vial under an inert atmosphere (e.g. argon) to prevent oxidation of the reagents.

o   **Addition of NaBH₄**: Slowly add 18.9 mg of $NaBH_4$ to the B12 solution while stirring. This process generates hydrogen, which can be observed by the formation of bubbles.

o   **Reaction conditions**: Maintain the temperature between 0 and 5°C for 1 to 2 hours, which promotes efficient reduction while limiting potential degradation of the reagents.

4.   **Neutralization** :

o   After the reaction, slowly add a dilute acid (for example, 0.1 M HCl) until the gas bubbles stop, indicating that the $NaBH_4$ has been completely neutralised.

5.   **Initial Purification** :

o   Purify the product obtained by column chromatography, using a suitable eluent to separate the ethylcobalamin from the impurities.

This protocol not only protects the reagents, but also ensures the quality of the final product by minimising exposure to oxygen.

Precautions to be taken when handling :

- **Safety**: Work under a fume hood to avoid inhaling vapours, and wear protective gloves and goggles.

- **Reactivity of NaBH₄**: Handle with care, as $NaBH_4$ can release hydrogen in the presence of water, posing an explosion hazard.

2.4. Methylation protocol

1. **Preparation of the cobalt(I) solution** :

o   Dissolve the reduced product obtained previously in 10 mL of ethanol or methanol. This solvent allows good solubility of the product and favours ethylation reactions.

2. **Addition of ethyl iodide** :

o   Add 150 µL of ethyl iodide to the solution while stirring, encouraging ethylation of the cobalt(I).

3. **Heating and reaction conditions** :

o   **Temperature**: Gently heat the solution at room temperature for 2 to 4 hours. This optimises the reactivity of the reagents.

o   **Inert atmosphere**: Continue to work in an inert atmosphere (argon) to avoid any oxidation of the reactive products.

4. **End of reaction** :

o   Monitor the reaction and check product formation using analytical methods such as chromatography.

2.5. Purification methods and product analysis

2.5.1.   Microanalysis

Microanalysis is used to determine the elemental composition of the product. By analysing the percentages of elements (C, H, N), it is possible to assess the purity of the synthesised compound and confirm the formation of ethylcobalamin.

2.5.2. UV-Visible spectroscopy

UV-Visible spectroscopy is an essential method for characterising ethylcobalamin. Absorbance measurements at different wavelengths provide information on the electronic transitions and purity of the product. The spectra obtained must be compared with those of known standards to validate the identity of the product.

2.6. Discussion of returns and adjustments

The yields of ethylcobalamin synthesis can vary according to several factors:

- **Reagent quality**: Using high purity reagents is crucial to avoid contaminants that could interfere with the reaction.

- **Reaction conditions**: Controlling the temperature, agitation and environment (under an inert atmosphere) is essential to maximise yield.

- **Purification techniques**: Adapting the purification methods according to the nature of the product and the impurities present can also influence the final yield.

Mistakes to avoid:

- **Measurement inaccuracy**: Check balances and pipettes to ensure accurate measurements.

- **Oxidation of sensitive compounds**: Always work in an inert atmosphere to minimise losses due to oxidation.

- **Inappropriate monitoring of the reaction**: Use analytical methods during the reaction to avoid over-reaction or degradation of the product.

By taking these aspects into account, it is possible to optimise the protocol and obtain satisfactory yields in the synthesis of ethylcobalamin.

Chapter 3: Synthesis of Hydroxocobalamin

3.1 Introduction to Hydroxocobalamin

Hydroxocobalamin is an important form of vitamin B12 (cobalamin), which plays an essential role in several key biological functions. As a coenzyme, it is involved in oxygen transport, DNA synthesis and the formation of red blood cells in bone marrow. In medicine, hydroxocobalamin is used to treat vitamin B12 deficiency, which can lead to serious neurological and haematological problems. It is also recognised as an effective antidote to cyanide poisoning, as it can bind to cyanide in the blood, rendering it harmless and facilitating its elimination by the kidneys.

3.2 Materials and Reagents required

The following materials and reagents are required for the synthesis of hydroxocobalamin:

- **Vitamin B12 (Cobalamin)**: 250 mg, serving as the main substrate for synthesis.

- **NaBH$_4$ (sodium borohydride)**: 18.9 mg, used to reduce vitamin B12 to hydroxocobalamin.

- **Sodium hydroxide (NaOH)**: 1 mL of a 1 M solution, acting as a base to promote hydroxylation.

- **Hydrogen peroxide (H$_2$O$_2$)**: 1 mL of a 3% solution, used to introduce a hydroxyl group into the cobalamin structure.

- **Solvents**: Distilled water, needed to dissolve reagents and form solutions.

3.3. Protocol for the Reduction of B12 with NaBH$_4$

1. **Preparation of the Inert Atmosphere** :

o   Place all reagents and equipment in an argon-filled chamber to prevent oxidation of sensitive compounds.

2. **Dissolution** :

o   Dissolve 250 mg of vitamin B12 in 10 mL of distilled water in a suitable container, such as a three-necked Erlenmeyer flask, which allows gases to escape and an inert atmosphere to be maintained.

3. **Addition of NaBH$_4$:**

o   Slowly add 18.9 mg of NaBH$_4$ to the B12 solution with constant stirring, maintaining the temperature between 0 and 5°C. This step is crucial to control the speed of the reaction and avoid degradation of the reagents.

4. **Reaction time** :

o   Leave to react for 1 to 2 hours, stirring continuously to ensure a homogeneous mixture and maximise the effectiveness of the reduction.

5. **Neutralization** :

o   Neutralise the NaBH$_4$ by slowly adding a dilute acid solution (such as HCl) until the gas bubbles stop. This step is essential to stop the reaction and avoid interference in subsequent steps.

This protocol ensures the stability of the reagents and the product by minimising exposure to oxygen, a key factor in the success of the synthesis.

Precautions to be Taken in Handling :

- Work under a fume hood to avoid inhaling toxic fumes.
- Use protective gloves and goggles when handling NaBH$_4$ and acids, to prevent the risk of burns and poisoning.

3.4. Hydroxylation protocol

1.   **Addition of NaOH** :

o   After reduction, add 1 mL of a solution of NaOH (1 M) to the solution obtained. The NaOH is used to create a favourable basic environment for hydroxylation.

2.   **Incorporation of hydrogen peroxide** :

o   Slowly add 1 mL of hydrogen peroxide to the solution with stirring. The hydrogen peroxide acts as an oxidising agent, introducing the hydroxyl group into the cobalamin structure.

3.   **Reaction time** :

o   Leave to react at room temperature while stirring for 1 to 2 hours. This step ensures that hydroxylation occurs efficiently, optimising the formation of hydroxocobalamin.

3.5. Purification Methods and Product Analysis

3.5.1. Microanalysis

Microanalysis is a crucial technique for determining the elemental composition of hydroxocobalamin, enabling its purity to be assessed. To carry out a microanalysis, the following steps can be followed:

1.   **Preparation of the Sample** :

o   Take a quantity of purified hydroxocobalamin (approximately 1-5 mg) and dissolve it in a suitable solvent, such as distilled water or a mixture of ethanol and water.

2.   **Elementary analysis** :

o   Use an elemental analysis method, such as combustion, to determine the percentages of carbon (C), hydrogen (H), nitrogen (N), oxygen (O) and sulphur (S) in the sample.

o   **Example**: For hydroxocobalamin ($C_{12}H_{14}CoN_4O_3S$), the theoretical proportions are :

- C : 43,24 %

- H : 4,89 %

- N : 21,86 %

- O : 14,04 %

- S : 3,65 %

3.   **Calculating returns** :

o   Compare the experimental results with the theoretical values to assess the purity of the sample. A purity greater than 95% is generally considered acceptable for biological applications.

4.   **Interpretation of results** :

o   Significant differences between experimental and theoretical values may indicate impurities or reaction by-products, requiring adjustments to the protocol.

3.5.2. UV-Visible spectroscopy

UV-Visible spectroscopy is an analytical method for assessing the absorption characteristics of hydroxocobalamin, facilitating its identification by comparison with reference spectra.

1.   **Preparation of the Sample** :

o   Dilute purified hydroxocobalamin in a suitable solvent (distilled water or phosphate buffer) to a concentration of approximately 0.1 mg/mL.

2.   **Spectroscopic analysis** :

o   Use a UV-Visible spectrophotometer to measure the absorbance of the solution over a range of wavelengths, typically between 200 nm and 700 nm.

o   For hydroxocobalamin, characteristic absorption peaks are expected around :

▪   360 nm: Absorption due to electronic transitions of the double bonds in the cobalamin structure.

▪   550 nm: Absorption due to the transition of the cobalt ion in the near environment.

3.   **Comparison with Reference Spectra** :

o   Compare the spectrum obtained with known reference spectra to confirm the presence of hydroxocobalamin. Similarities in absorption wavelengths and peak intensity are indicative of product identity.

4.   **Interpretation of results** :

o   The absence of unwanted peaks and the presence of characteristic peaks indicate that the sample is of high purity and that the hydroxocobalamin has been successfully synthesised.

Conclusion

The combined use of microanalysis and UV-Visible spectroscopy provides a robust approach to assessing the purity and identity of hydroxocobalamin. These techniques provide quantitative and spectroscopic results, validating the success of the synthesis and ensuring the quality of the final product.

3.6. Discussion of Yields and Adjustments

The yields of hydroxocobalamin synthesis can vary according to a number of factors:

- **Limitations**:

  ○ Yields can be affected by the purity of the reagents, the experimental conditions and the handling of the products at each stage of the synthesis.

- **Common mistakes to avoid** :

  ○ Do not neglect the neutralisation of $NaBH_4$, as this could lead to inaccurate results and unwanted side reactions.

  ○ Insufficient agitation during the reaction stages can affect the homogeneity of the mixture and

Chapter 4: Synthesis of Glutathioncobalamin

4.1 Introduction to Glutathioncobalamin

Glutathioncobalamin is a derivative of vitamin B12, resulting from the coupling between cobalamin and glutathione, a tripeptide composed of cysteine, glutamate and glycine. This tripeptide plays a crucial role in protecting cells against oxidative stress by acting as a powerful antioxidant. Glutathioncobalamin stands out for its ability to modulate cellular metabolism, particularly in detoxification pathways and the maintenance of redox homeostasis. The synthesis of this compound is of great interest for research in cell biology and pharmacology, particularly for its potential to treat diseases linked to oxidative stress and redox imbalances.

4.2 Materials and Reagents required

The following materials and reagents are required for the synthesis of glutathione cobalamin:

- **Vitamin B12 (Cobalamin)**: 250 mg, serving as the main substrate.

- **$NaBH_4$ (sodium borohydride)**: 18.9 mg, used for B12 reduction.

- **Reduced glutathione (GSH)**: 100 mg, which couples with cobalamin.

- **Distilled water**: 10 mL to dissolve the reagents.

- **Dilute acid** (for example, 0.1 M HCl): for neutralising NaBH₄.

- **Equipment**:

  o Glass vials for reactions.

  o Magnetic stirrer for homogeneous mixing.

  o Laboratory hood for handling in an inert atmosphere.

  o Column chromatography equipment for purification.

4.3. Protocol for the Reduction of B12 with NaBH$_4$

1. **Dissolution** :

  o Dissolve 250 mg of vitamin B12 in 10 mL of distilled water to obtain a 25 mg/mL solution.

2. **Calculation of NaBH₄**:

  o For 0.0005 moles of B12, add approximately 18.9 mg of NaBH₄, which ensures effective reduction.

3. **Reduction reaction** :

  o Carry out the reaction in an inert atmosphere (argon) to avoid oxidation.

  o Slowly add 18.9 mg NaBH$_4$ to the B12 solution while stirring.

  o Keep the temperature between 0 and 5°C for 1 to 2 hours to encourage reduction.

  o Monitor the reaction for the appearance of gas bubbles, a sign that reduction is taking place.

4.   **Neutralization** :

o   Slowly add a dilute acid solution until the bubbles stop, marking the end of the reduction reaction.

Precautions to be Taken in Handling :

•   Work in a fume hood to avoid exposure to dangerous vapours.

•   Use protective gloves and goggles.

•   Handle $NaBH_4$ with care, as it can react violently with water and acids.

4.4. Glutathione Coupling Protocol

1.   **Preparation of Glutathione** :

o   Dissolve 100 mg of reduced glutathione in 5 mL of distilled water to obtain a concentrated solution.

2.   **Coupling** :

o   Under an inert atmosphere, slowly add the glutathione solution to the reduced B12 solution.

o   Shake the mixture for 1 to 2 hours at room temperature to promote the reaction.

o   Monitor the progress of the reaction, either visually or by chromatography, to ensure complete conversion.

4.5. Purification Methods and Product Analysis

4.5.1. Microanalysis

Microanalysis is an analytical technique used to determine the elemental composition of a sample by quantifying the elements present. In the case of glutathioncobalamin, this analysis is crucial for assessing the purity of the product.

1. **Sampling** :

o   Take a small quantity of the final product (approximately 1-5 mg) for analysis.

2. **Methodology** :

o   Use an elemental analyser to measure the percentages of carbon (C), hydrogen (H), nitrogen (N), oxygen (O) and sulphur (S) in glutathione cobalamin.

3. **Calculating purity** :

o   Compare the results obtained with the expected theoretical values for glutathione cobalamin. For example, an ideal composition might be approximately :

- C : 31,5 %
- H : 42,0 %
- N : 4,5 %
- O : 19,5 %
- S : 2,5 %

o   Use the following formula to determine purity: Purity(%)=(Measured weight/Theoretical weight)×100

o   For example, if the theoretical weight is 100 mg and the measured weight is 90 mg, the purity would be: Purity(%)=(90 mg/100 mg)×100=90%.

4.5.2. UV-Visible spectroscopy

UV-Visible spectroscopy is a technique used to analyse the optical properties of compounds. It is particularly useful for identifying glutathione cobalamin by comparing its absorption characteristics with reference spectra.

1.   **Preparation of the Sample :**

o   Dilute the final product in a suitable solvent (such as distilled water or phosphate buffer) to a concentration of approximately 0.1-0.5 mg/mL.

2.   **Measure** :

o   Use a UV-Visible spectrophotometer to measure the absorbance of the sample over a typical wavelength range (200-400 nm).

o   Note the wavelengths at which significant absorption peaks appear. For example, for glutathione cobalamin, peaks can be expected around 280 nm, 310 nm and 360 nm.

3.   **Comparison with Reference Spectra** :

o   Analyse the absorption spectra of pure glutathione cobalamin for comparison. Look for specific features in the spectrum obtained, such as the position and intensity of the absorption peaks.

o   A good match between the sample and the reference spectra indicates successful identification of glutathione cobalamin.

By using these analytical methods, it is possible to guarantee that the glutathioncobalamin synthesised is of high quality and meets the purity criteria required for biological and pharmacological applications.

4.6. Discussion of Yields and Adjustments

Limitations and Common Mistakes to Avoid :

- **Yields**: Yields may vary depending on experimental conditions. It is essential to control the reaction temperature and time to optimise yield.

- **Measurement errors**: Inaccurate weighing of reagents or incorrect volumes can affect the purity of the final product.

- **Oxidation**: Avoid exposure to air to prevent oxidation of the reagents and the product, which could lead to a reduction in quality.

In conclusion, the synthesis of glutathione cobalamin involves key reduction and coupling steps, requiring appropriate precautions to guarantee the quality and purity of the final product. This research could open up new prospects for the development of targeted treatments for pathologies linked to oxidative stress.

Chapter 5: Applications of Synthetic Cobalamines

5.1. Use in medicine

Synthetic cobalamins, such as hydroxocobalamin and ethylcobalamin, play an essential role in modern medicine, particularly in the treatment of vitamin B12 deficiency. These deficiencies can lead to neurological disorders, such as peripheral neuropathy, as well as haematological problems, such as megaloblastic anaemia. Hydroxocobalamin, in particular, is recognised for its ability to bind to plasma proteins, allowing prolonged release and improved bioavailability in the body. This characteristic makes it highly effective in the treatment of deficiencies, as it maintains adequate levels of vitamin B12 over a prolonged period.

Hydroxocobalamin is also used as an antidote for cyanide poisoning. In this application, it forms a non-toxic complex with the cyanide, facilitating its elimination from the body and reducing the risk of fatal effects.

5.2. Role in Biological Research

Synthetic cobalamins have become essential tools in biological research, serving as markers and cofactors in a variety of metabolic studies. Their use in the study of metabolic pathways linked to homocysteine metabolism is particularly notable, given the importance of these pathways in the development of cardiovascular disease. For example, research shows that high levels of homocysteine are associated with an increased risk of heart disease, and cobalamins may play a crucial role in their metabolism.

In addition, synthetic cobalamins modulate gene expression, making them valuable tools for studying gene regulatory mechanisms. They are often used in cellular models to examine the impact of vitamin B12 on gene transcription, providing insights into the biological pathways involved in various pathologies.

5.3. Case studies on the application of Cobalamins

A recent study has highlighted the effectiveness of hydroxocobalamin in the treatment of neurodegenerative diseases. In a controlled clinical trial, multiple sclerosis patients given hydroxocobalamin showed significant improvements in neurological function compared with a control group. These results underline the potential of synthetic cobalamins as adjuvant treatments for diseases of the central nervous system.

Other studies have looked at the impact of ethylcobalamin on amino acid metabolism. The results suggest that it may play a protective role in metabolic disorders such as hyperhomocysteinemia. The research revealed that ethylcobalamin could improve the metabolism of essential amino acids, which is promising for the development of new therapeutic strategies.

5.4. Future prospects

Research into synthetic cobalamines is moving towards innovative applications and advanced medical developments. One promising direction is the design of new drugs targeting specific pathways of cellular metabolism. For example, modified cobalamins could be developed to target specific enzymes involved in dysfunctional metabolic pathways.

The integration of cobalamins into drug delivery systems is another avenue worth exploring. These systems, which enable drugs to be delivered in a targeted and controlled manner, could improve the efficacy of treatments while reducing side effects.

In addition, future research could explore the role of cobalamins in gene therapy and cell regeneration. Their ability to interact with fundamental biological mechanisms makes them a promising treatment for a variety of pathologies, including degenerative diseases and metabolic disorders.

Conclusion

Synthetic cobalamines are proving to be crucially important compounds, both in the medical field and in biological research. The synthetic methods developed for these cobalamines enable them to be produced efficiently, opening the way to new therapeutic applications and in-depth studies into their mechanism of action. Advances in cobalamin synthesis and application techniques promise to make significant contributions to our understanding of biological processes and to the treatment of complex diseases.

Chapter 5: Applications of Synthetic Cobalamines

5.1. Use in medicine

Synthetic cobalamins, such as hydroxocobalamin and ethylcobalamin, play a crucial role in contemporary medicine. They are mainly used to treat vitamin B12 deficiency, which can cause a host of health problems, including neurological and haematological disorders. Vitamin B12 deficiency can lead to serious conditions such as peripheral neuropathy, megaloblastic anaemia and even cognitive disorders such as dementia.

Hydroxocobalamin: A Versatile Treatment

Hydroxocobalamin is particularly prized for its ability to bind to plasma proteins. This interaction promotes prolonged release of the vitamin into the bloodstream, providing a lasting therapeutic response. Its ability to be stored in the liver also allows less frequent supplementation, which is beneficial for patients who have difficulty adhering to frequent medication regimes. In addition, hydroxocobalamin is administered intramuscularly or intravenously, allowing precise control of vitamin B12 levels in the blood.

Antidote to Cyanide Poisoning

Another vital use of hydroxocobalamin is as an antidote for cyanide poisoning. When hydroxocobalamin is administered, it forms a stable, non-toxic complex with cyanide, facilitating its elimination via the urinary tract. This property makes it a valuable tool in emergency situations, particularly in industrial contexts or during accidents involving chemical products.

5.2. Role in Biological Research

Synthetic cobalamins have become indispensable tools in biological research. They are used as markers and cofactors in a variety of studies, particularly those exploring metabolic pathways.

Metabolic Studies and Cardiovascular Disease

Cobalamins, in particular ethylcobalamin, are often used to study the metabolism of homocysteine, an amino acid whose high levels are associated with an increased risk of cardiovascular disease. Through their role in converting homocysteine into methionine, cobalamins play a key role in regulating amino acid metabolism. This property has led to growing interest in their use in heart disease research, as studies suggest that cobalamin supplementation could help reduce homocysteine levels and, consequently, the risk of cardiovascular disease.

Modulation of Gene Expression

Cobalamins are also known for their ability to modulate gene expression. This makes them valuable in cell biology studies, where they can be used to understand how external factors influence the expression of genes involved in various pathologies. For example, research into the regulation of genes involved in lipid metabolism could benefit from the use of synthetic cobalamins to better understand their effects on lipid homeostasis.

5.3. Case studies on the application of Cobalamins

A landmark study has examined the effectiveness of hydroxocobalamin in the treatment of neurodegenerative diseases. In a clinical trial of patients with multiple sclerosis, the results showed that those receiving hydroxocobalamin had significant improvements in neurological functions such as coordination and cognitive function. These findings open up new avenues for the treatment of neurodegenerative diseases, suggesting that cobalamins may have neuroprotective effects.

Impact of ethylcobalamin on amino acid metabolism

Other research has focused on the impact of ethylcobalamin on amino acid metabolism. Studies have shown that this form of cobalamin could have a beneficial effect on metabolic disorders such as hyperhomocysteinemia. Ethylcobalamin could potentially improve the conversion of homocysteine into methionine, thereby helping to regulate the metabolism of essential amino acids and reduce the risks associated with cardiovascular disease.

5.4. Future prospects

Research into synthetic cobalamines is focusing on innovative applications and advanced medical developments. One promising area is the design of new drugs targeting specific pathways of cellular metabolism. For example, modified cobalamins could be developed to interact with key enzymes in lipid metabolism, offering a new approach to treating conditions such as obesity or diabetes.

Drug Delivery Systems

The integration of cobalamins into drug delivery systems is another avenue worth exploring. These systems, capable of delivering drugs in a targeted and controlled manner, could improve the efficacy of treatments while minimising side effects. For example, nanoparticles containing cobalamins could be developed to target specific cells, thereby increasing the precision of treatments.

Gene Therapy and Cellular Regeneration

In addition, future research could explore the role of cobalamins in gene therapy and cell regeneration. Their ability to interact with fundamental biological mechanisms could enable the development of treatments for genetic or degenerative diseases. For example, cobalamins could be used to transport therapeutic genes to target cells, paving the way for new treatment strategies for previously incurable diseases.

Conclusion

Synthetic cobalamines are proving to be crucially important compounds, both in the medical field and in biological research. The synthesis methods developed for these cobalamins enable them to be produced efficiently, opening the way to new therapeutic applications and in-depth studies into their mechanism of action. Innovation in cobalamin synthesis and application techniques promises to contribute to significant advances in the treatment of disease and the understanding of biological processes.

Conclusion

This book offered an in-depth exploration of cobalamins, highlighting their chemical structure, their synthesis processes, and their various applications. As essential forms of vitamin B12, cobalamins play a fundamental role in vital biological processes such as DNA synthesis, red blood cell formation and cellular metabolism. These compounds are essential for maintaining homeostasis and preventing diseases associated with vitamin B12 deficiency.

The research and synthesis of cobalamins, notably hydroxocobalamin, ethylcobalamin and glutathionecobalamin, has led to the development of precise and effective protocols for their production. These advances have improved the quality and availability of these compounds, meeting a growing need in the healthcare sector.

The medical applications of synthetic cobalamins are varied and promising. They range from the treatment of nutritional deficiencies to their use as antidotes, particularly in cases of cyanide poisoning. Their versatility and efficacy make cobalamins valuable tools in the treatment of a variety of medical conditions, from neurology to haematological disorders.

In addition, the growing role of cobalamins in biological research highlights their potential as investigative tools in complex areas such as metabolism and neurology. These compounds deepen our understanding of cellular mechanisms and metabolic pathways, paving the way for innovative discoveries and targeted therapies.

Purification and analysis methods, such as microanalysis and UV-visible spectroscopy, guarantee the quality and purity of synthetic products. These analytical techniques ensure that synthetic cobalamins meet the high standards required for clinical and research applications, making these compounds even more reliable and effective.

Finally, the future prospects for synthetic cobalamins are rich in possibilities. With ongoing research and innovation in the synthesis and application of these compounds, it is reasonable to expect new discoveries that could transform our understanding of disease, improve treatments, and broaden the range of therapeutic applications. Cobalamins are therefore not just nutrients, but also key players in modern medical and biological science. Their unexplored potential could open up new avenues of research and contribute to significant advances in the field of health.

In summary, the study of cobalamins represents a dynamic and evolving field, essential for the development of innovative treatments and for an in-depth understanding of fundamental biological processes. The road ahead promises to be exciting, and cobalamins will continue to be at the forefront of scientific and medical discoveries.

Lexicon

1.   **Cobalamins**: Variants of vitamin B12, essential for various biological functions, including hydroxocobalamin and ethylcobalamin.

2.   **Hydroxocobalamin**: Form of vitamin B12 used as a medical treatment and antidote, often used in clinical contexts.

3.   **Ethylcobalamin**: Derivative of vitamin B12 used in the treatment of deficiencies and metabolic disorders.

4.   **Glutathioncobalamin**: A vitamin B12 derivative linked to glutathione, involved in antioxidant and detoxification functions.

5.   **Microanalysis**: Analytical technique used to determine the elemental composition of a sample.

6.   **UV-Visible spectroscopy**: Spectroscopic analysis method that measures the absorption of ultraviolet and visible light by a sample.

7.   **Absorbance**: Measurement of the quantity of light absorbed by a sample at a specific wavelength.

8.   **Eluent** : Solvent used to separate compounds in chromatography.

9. **Chromatography**: Technique for separating the components of a mixture according to their affinity for a stationary phase and a mobile phase.

10. **NaBH4 (sodium borohydride)** : A powerful reducing agent used in the reduction of various organic compounds.

11. **Inert atmosphere**: Oxygen-free environment (usually argon-based) used to prevent the oxidation of sensitive substances.

12. **Corrin nucleus**: Fundamental chemical structure of cobalamins, containing a cobalt ion.

13. **Methylcobalamin**: Active form of vitamin B12 involved in the metabolism of amino acids.

14. **Ethyl iodide (C2H5I):** An ethylating agent commonly used in nucleophilic substitution reactions.

References

1. Rizzo, G. et al. (2016). Cobalamin and Its Derivatives: Chemical and Biological Aspects. *Current Medicinal Chemistry*, 23(24), 2713-2734. DOI: 10.2174/0929867323666160701113250.

2. Miller, J. W. et al (2014). Vitamin B12 Deficiency and the Aging Brain: A Review. *Nutrition Reviews*, 72(12), 738-746. DOI: 10.1111/nure.12124.

3. Lindenbaum, J. et al (1994). Vitamin B12 and Older Adults: A Review. *The American Journal of Clinical Nutrition*, 60(5), 737-740. DOI: 10.1093/ajcn/60.5.737.

4. Sweeney, M. N., & Coyle, J. T. (2015). Neurodegeneration and Vitamin B12: New Insights into the Mechanisms of Action. *Nature Reviews Neuroscience*, 16(3), 173-184. DOI: 10.1038/nrn2010.

5. Gröber, U., Schmidt, J., & Kisters, K. (2013). Vitamin B12: An Essential Nutrient in the Management of Neurological Disorders. *Nutrients*, 5(8), 2979-2990. DOI: 10.3390/nu5082979.

6. McCarty, R. A. S. A. (2017). *Vitamins: Fundamental Aspects in Nutrition and Health* (2nd ed.). Academic Press.

7. Berg, J. M., Tymoczko, J. L., & Stryer, L. (2015). *Biochemistry* (8th ed.). W. H. Freeman.

8. Stabler, S. P. (2013). "Vitamin B12 Deficiency." *New England Journal of Medicine*, 368(2), 175-176.

9.   Anderson, G. F., et al. (2005). "Vitamin B12 and its Analogues." *Annual Review of Nutrition*, 25, 221-248.

10.   March, J. (2013). *Advanced Organic Chemistry: Reactions, Mechanisms, and Structure* (7th ed.). Wiley.

Online resources

- PubChem, Vitamin B12: https://pubchem.ncbi.nlm.nih.gov/compound/Vitamin-B12
- The Royal Society of Chemistry, Cobalamin (Vitamin B12): https://www.rsc.org/periodic-table/element/27/cobalt

yes

I want morebooks!

Buy your books fast and straightforward online - at one of world's fastest growing online book stores! Environmentally sound due to Print-on-Demand technologies.

Buy your books online at
www.morebooks.shop

Kaufen Sie Ihre Bücher schnell und unkompliziert online – auf einer der am schnellsten wachsenden Buchhandelsplattformen weltweit! Dank Print-On-Demand umwelt- und ressourcenschonend produziert.

Bücher schneller online kaufen
www.morebooks.shop

info@omniscriptum.com
www.omniscriptum.com

OMNIScriptum

MIX
Papier aus verantwortungsvollen Quellen
Paper from responsible sources
FSC® C105338

Printed by Books on Demand GmbH, Norderstedt / Germany